Collins
INTERNATIONAL

T0173336

Maths
Foundation

Activity Book A

Published by Collins
An imprint of HarperCollins*Publishers*
The News Building, 1 London Bridge Street,
London, SE1 9GF, UK

HarperCollins*Publishers*
Macken House, 39/40 Mayor Street Upper,
Dublin 1, D01 C9W8, Ireland

Browse the complete Collins catalogue at
www.collins.co.uk

© HarperCollins*Publishers* Limited 2021

10 9 8 7 6 5 4 3

ISBN 978-0-00-846877-4

British Library Cataloguing-in-Publication Data
A catalogue record for this publication is available from the British Library.

Author: Peter Clarke
Publisher: Elaine Higgleton
Product manager: Letitia Luff
Commissioning editor: Rachel Houghton
Edited by: Sally Hillyer
Editorial management: Oriel Square
Cover designer: Kevin Robbins
Cover illustrations: Jouve India Pvt Ltd.
Internal illustrations: Jouve India Pvt. Ltd.
Typesetter: Jouve India Pvt. Ltd.
Production controller: Lyndsey Rogers
Printed and bound in India by
Replika Press Pvt. Ltd.

Acknowledgements

With thanks to all the kindergarten staff and their schools around the world who
have helped with the development of this course, by sharing insights and
commenting on and testing sample materials:

Calcutta International School: Sharmila Majumdar, Mrs Pratima Nayar, Preeti
Roychoudhury, Tinku Yadav, Lakshmi Khanna, Mousumi Guha, Radhika Dhanuka,
Archana Tiwari, Urmita Das; Gateway College (Sri Lanka): Kousala Benedict; Hawar
International School: Kareen Barakat, Shahla Mohammed, Jennah Hussain; Manthan
International School: Shalini Reddy; Monterey Pre-Primary: Adina Oram; Prometheus
School: Aneesha Sahni, Deepa Nanda; Pragyanam School: Monika Sachdev; Rosary
Sisters High School: Samar Sabat, Sireen Freij, Hiba Mousa; Solitaire Global School:
Devi Nimmagadda; United Charter Schools (UCS): Tabassum Murtaza; Vietnam
Australia International School: Holly Simpson

The publishers wish to thank the following for permission to reproduce photographs.

(t = top, c = centre, b = bottom, r = right, l = left)

p 10tl michaeljung/Shutterstock, p 10tr polkadot_photo/Shutterstock,
p 10bl Firma V/Shutterstock, p 10br pixelheadphoto digitalskillet/Shutterstock,
p 11tl ESB Professional/Shutterstock, p 11tr michaeljung/Shutterstock,
p 11bl fizkes/Shutterstock, p 11br Andy Dean Photography/Shutterstock

MIX
Paper | Supporting
responsible forestry
FSC™ C007454

This book is produced from independently certified FSC™
paper to ensure responsible forest management.

For more information visit:
www.harpercollins.co.uk/green

Count and match

Draw lines to match the hands showing
the same number of fingers.

Date:

Count and draw

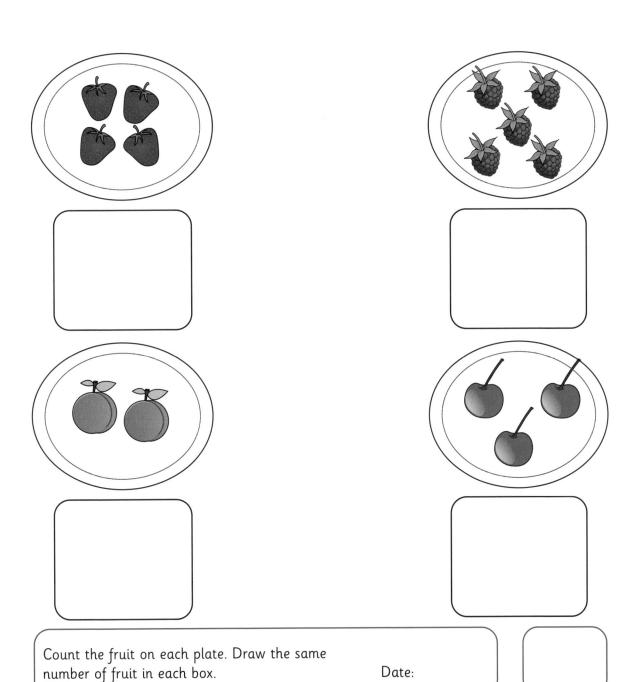

Count the fruit on each plate. Draw the same number of fruit in each box.   Date:

Listen and count

An adult will do these actions. Count how many.
Circle the matching number of pictures.　Date:

Count and draw

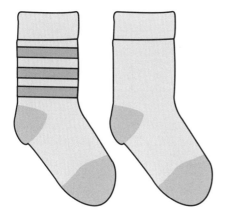

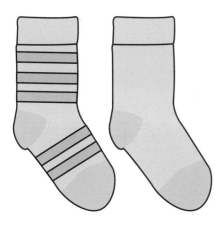

Count the stripes on the socks. Draw the same number
of stripes on the matching sock.   Date:

Count and match

1

2

3

4

5

Count the fingers on each hand. Draw a line to the matching number.

Date:

Trace and write

Trace the numbers. At the end of each row,
write the number once more.   Date:

Trace and circle

Trace number 1. Circle number 1 and 1 child. Trace number 2.
Circle number 2 and 2 children. Repeat for 4 and 5. 3 has
been done as an example.   Date:

More

For each pair, circle the photo that shows more people.   Date:

Position

Draw a toy **next to** the doll. Draw a toy **above** the rocket. Draw a toy **below** the drum.

Date:

Direction

For the top pictures, circle the children who are moving **up**. For the bottom pictures, circle the children who are moving **forwards**.   Date:

Near or far

Draw a frog **near** the duck. Draw a bird **far** from the tree. Draw a child **near** the bench.

Date:

Slide or roll

Circle in **blue** the objects that **slide**.
Circle in **red** the objects that **roll**.

Date:

Bigger

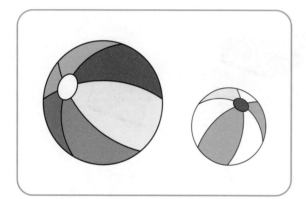

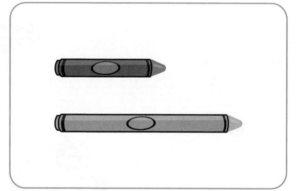

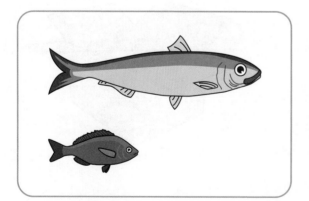

In each pair, circle the bigger object.   Date:

Longer

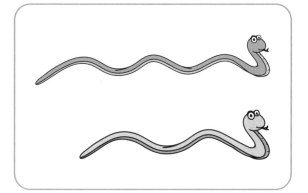

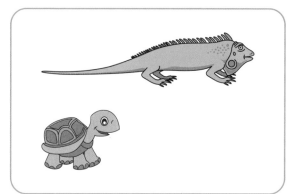

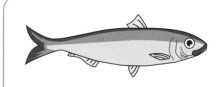

In each pair, circle the longer animal. Draw a longer fish.   Date:

Taller

In each pair, circle the taller animal. Draw a taller tree.   Date:

Wider

In each pair, circle the wider object. Draw a wider bowl.   Date:

Circles and triangles

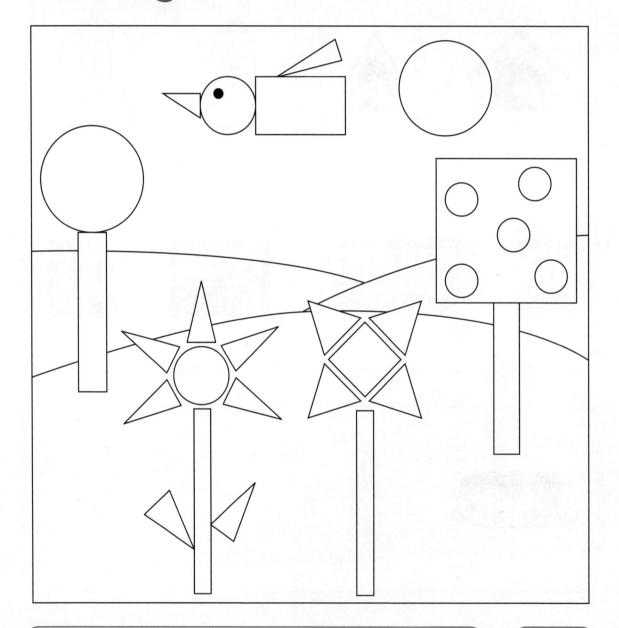

Colour all the circles yellow or red. Colour all the triangles purple or green.

Date:

Squares and rectangles

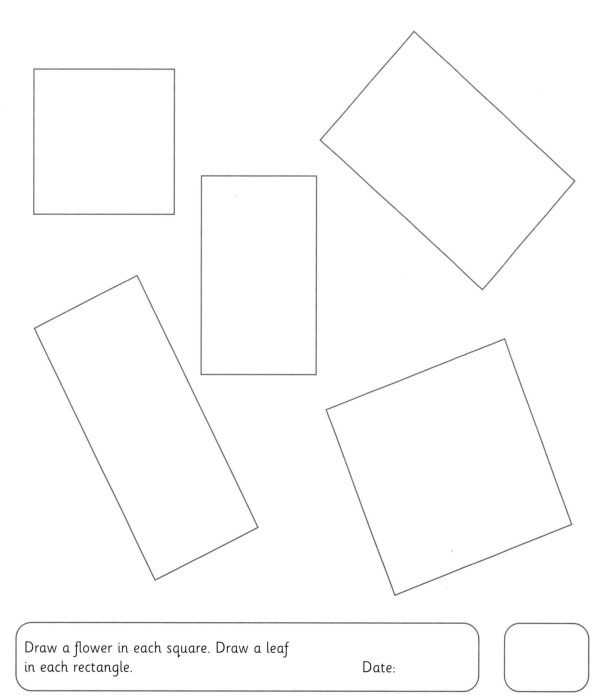

Draw a flower in each square. Draw a leaf in each rectangle.

Date:

Sort

Colour each shape to match the shapes
at the top of the page.   Date:

Shape patterns

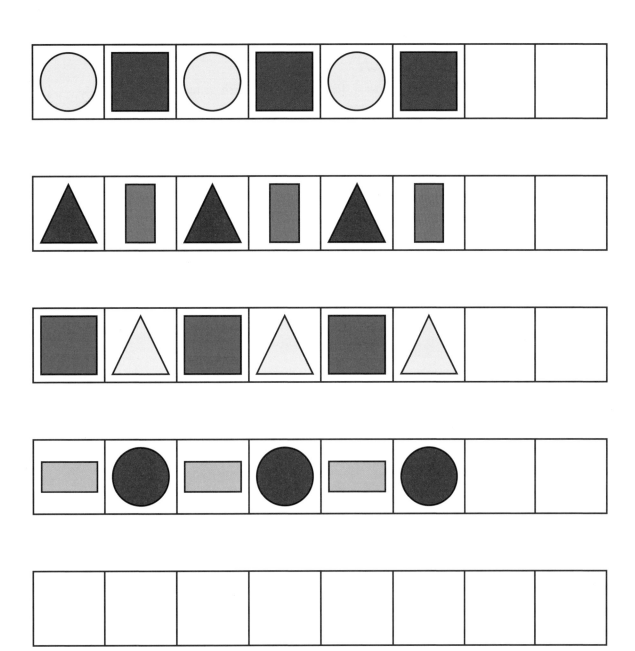

Draw the next two shapes in each pattern.
Draw your own repeating pattern.   Date:

Assessment record

_____ has achieved these Maths Foundation Phase Objectives:

Counting and understanding numbers

- Say and use the number names in order in familiar contexts such as number rhymes, songs, stories, counting games and activities, from 1 to 5.   1   2   3
- Say the number names in order, continuing the count on or back, from 1 to 5.   1   2   3
- Count objects from 1 to 5.   1   2   3
- Count in other contexts such as sounds or actions from 1 to 5.   1   2   3

Reading and writing numbers

- Recognise numbers from 1 to 5.   1   2   3
- Begin to record numbers, initially by making marks, progressing to writing numbers from 1 to 5.   1   2   3

Comparing and ordering numbers

- Use language such as more, less or fewer to compare two numbers or quantities from 1 to 5.   1   2   3

Patterns and sequences

- Talk about, recognise and make simple patterns using real-world objects or visual representations.   1   2   3

Understanding shape

- Identify, describe, compare and sort 2D shapes.   1   2   3

Position, direction and movement

- Begin to understand and use the vocabulary of position, direction and movement.   1   2   3

Measurement

- Use everyday language to describe and compare length, height and width, including long, longer, short, shorter, tall, taller, wide, wider, narrow and narrower.   1   2   3

Statistics

- Sort, represent and describe data using real-world objects or visual representations.   1   2   3

1: Partially achieved   2: Achieved   3: Exceeded

Signed by teacher:
Signed by parent: Date: